D1523446

C O C K R O A C H E S

Published by Smart Apple Media

123 South Broad Street

Mankato, Minnesota 56001

Printed in the United States of America.

Photos: James Castner; Dan L. Perlman (pages 6, 14);

PhotoDisc (page 24)

Design &Production: EvansDay Design

Project management: Odyssey Books

Library of Congress Cataloging-in-Publication Data

Richardson, Adele, 1966–

Cockroaches / Adele Richardson. – 1st ed.

p. cm. – (Bugs)

Includes bibliographical references and index.

Summary: Describes the habitat, life cycle, behavior,

predators, and unique characteristics of cockroaches.

ISBN 1-887068-31-7

1. Cockroaches—Juvenile literature. [1. Cockroaches.]

I. Title. II Series: Bugs (Mankato, Minn.)

QL505.5.R535 1998

595.7'28—dc21 *98-15346*

First Edition *9 8 7 6 5 4 3 2 1*

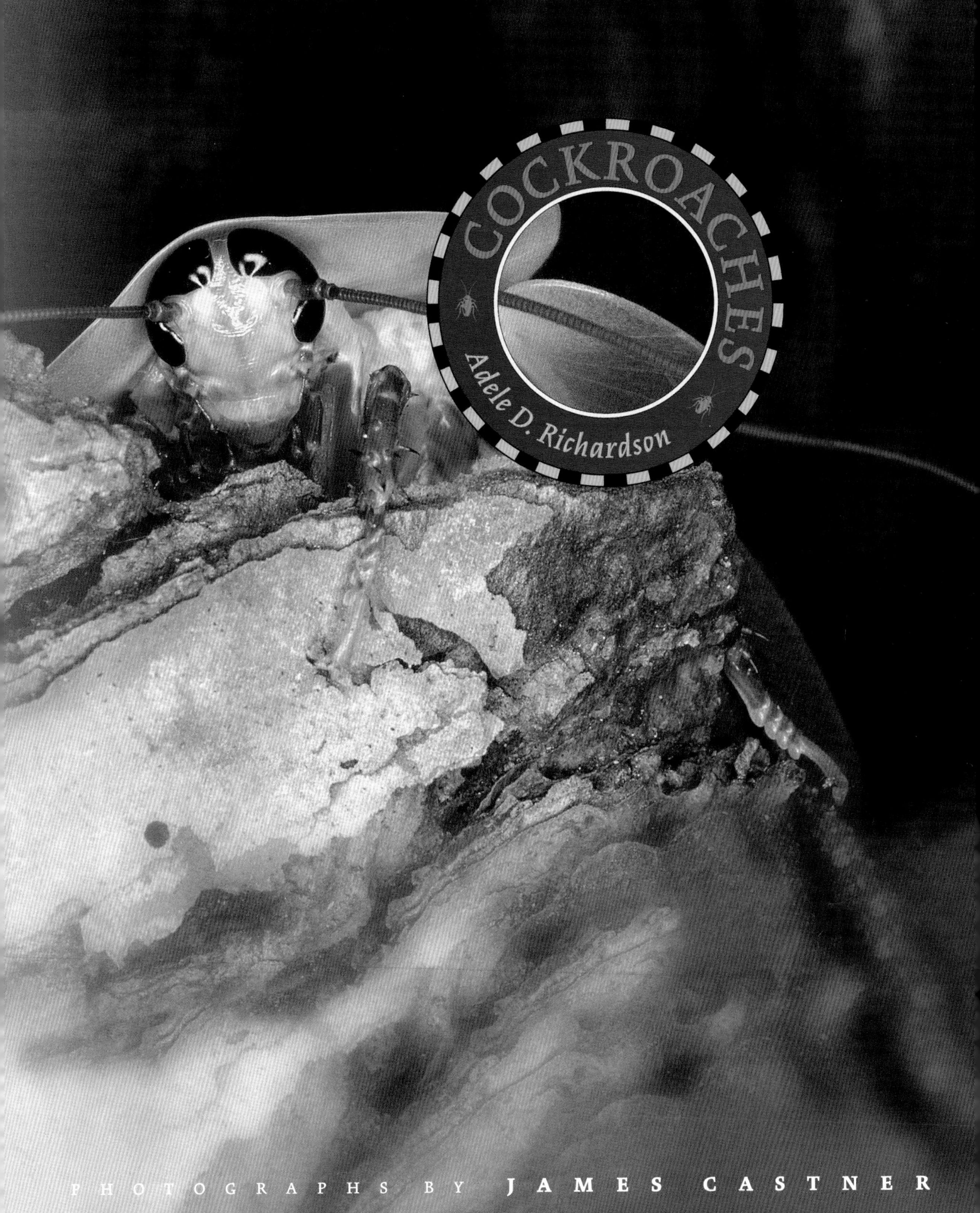
COCKROACHES
Adele D. Richardson
PHOTOGRAPHS BY JAMES CASTNER

It's nighttime and the family is tucked safely away in bed. In the kitchen is the *soft skittering* of tiny feet as an insect **RACES** across a darkened floor. It finds a bread crumb under the stove and settles down for a snack. Soon, more and more insects appear and ***DART*** across the room to find for themselves a tasty treat. Before you know it, the kitchen is infested! **Yuck!** It's *COCKROACHES!* Where did they all come from, and

why are there so many?

About Cockroaches

There are over 4,000 SPECIES, or kinds, of cockroaches in the world. Surprisingly, fewer than 1 percent (or 25 to 30 species) are considered to be pests. Of course, that really doesn't matter. If you spot a cockroach in the middle of the night, there's a good chance you will want to squish it.

Even worse news for those of you who are roach-haters: Scientists believe there may be thousands more species out there that haven't even been named yet!

Crazy Bug? Scientifically, cockroaches belong to the order *Blattaria*. The name

Long antennae, like these on the American cockroach, are the insect's most important sense organ.

means "to shun the light." Their common name, cockroaches, comes from the Spanish word *cucaracha*, which means "crazy bug." Anyone who has ever seen a cockroach zigzag all over the place can understand how it came by that name.

Warm, Damp Homes Cockroaches live around the world in warm, moist environments. That's why they like basements, bathrooms, and kitchens so well. Humans like to stay warm, and where there are humans there is usually food and

This Tropical cockroach is named after the climate it likes to live in.

water too. So naturally, cockroaches want to move in with us.

Even though these bugs seek warm homes, some have even been found at the North and South Poles. They were accidentally brought there by humans and live hidden away in the cracks and crevices of the man-made shelters.

Roaches Among Us

In the United States there are 57 known species—32 species are found in Texas alone! They range in size from about 1/2 inch to 2 inches long (1.3 to 5 cm), and most are brown or black in color. Cockroaches can fly and swim as well as run fast. Their bodies even have a waxy covering that helps to keep them from drowning. No wonder they're so hard to catch!

Roach Bodies

Like many other insects, cockroaches have an EXOSKELETON, or outer covering, which provides body support and protection. Along with all other insects, they have six legs and breathe air. Their bodies, whether big or small, are divided into three sections: head, thorax, and abdomen.

Head One of the most important features, and often the most recognizable, on

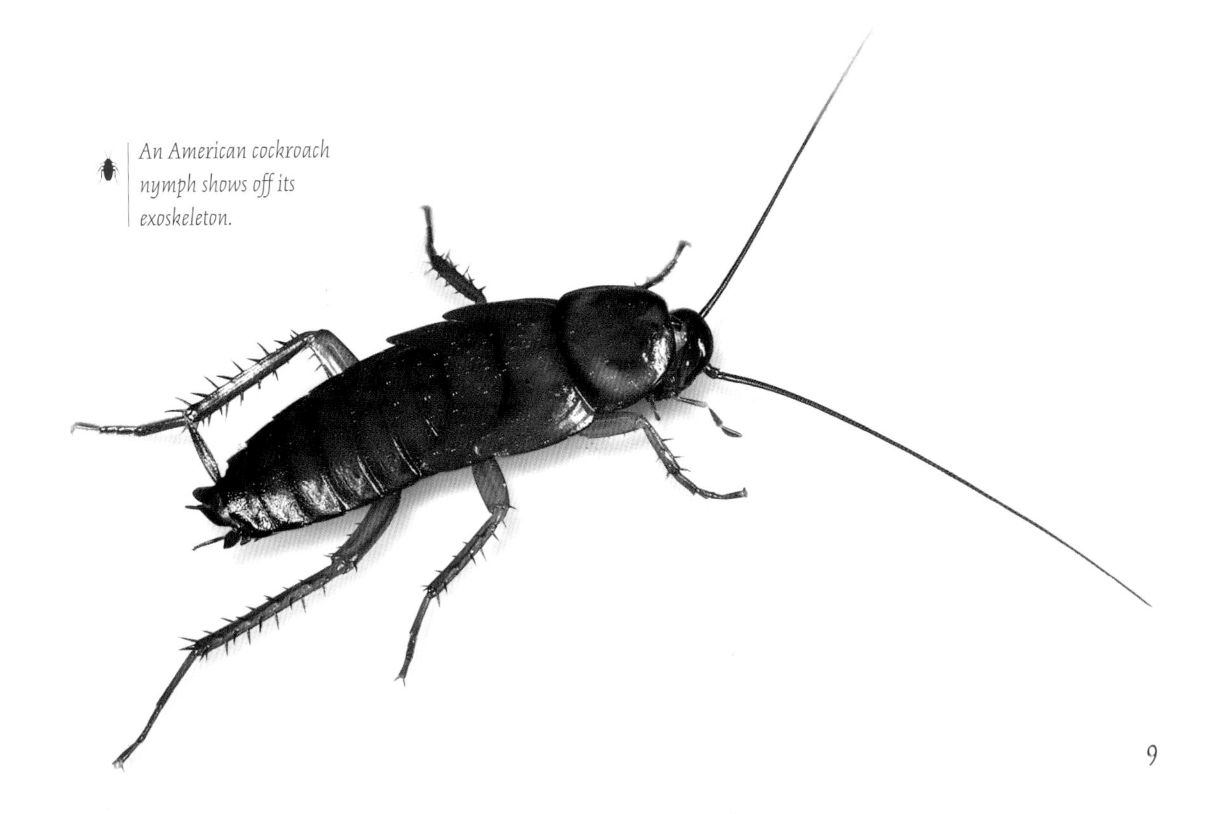

An American cockroach nymph shows off its exoskeleton.

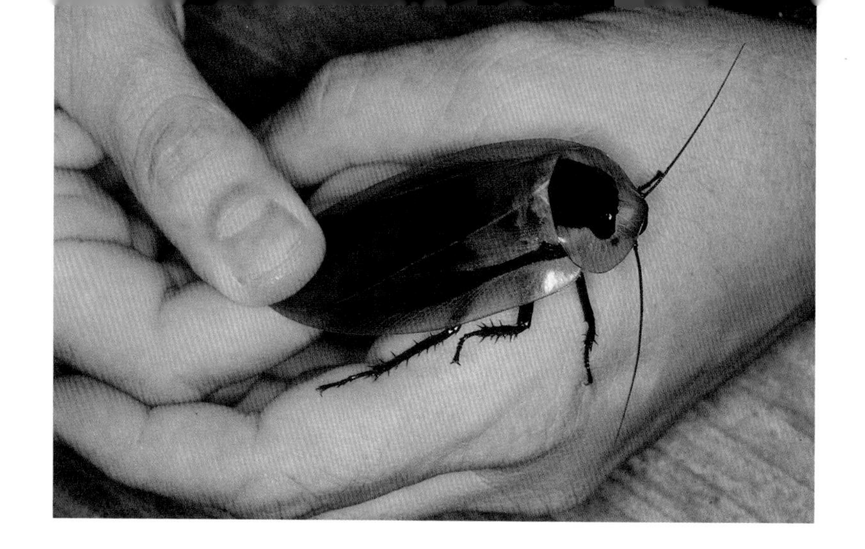

a cockroach is the ANTENNAE, or feelers. The antennae are usually longer than the insect's entire body. They are used to taste, smell, and feel as well as to identify moisture and the direction in which the insect is heading. Its antennae are constantly moving, even when the cockroach is resting.

The male cockroach uses his antennae to detect special chemicals called PHEROMONES that the female emits when it's time to mate. In truth, a cockroach's antennae are even more important than its eyes.

Cockroach Sees All The cockroach's eyes, called COMPOUND EYES, are set on the top of its head. They are made up of many small lenses that help the insect detect

An adult Giant cockroach (Blaberus giganteus) *is almost as big as a human thumb.*

movement in almost any direction. We humans have only one lens in each eye.

The mouth is also located on the head. It contains many parts and is used to smell as well as taste. When chewing, the insect's MANDIBLES, or jaws, move from side to side instead of up and down like ours. Another mouthpart is the tongue,

The mouthparts of a newly molted Giant cockroach are difficult to see.

called a HYPOPHARYNX, which the insect uses to lick food.

A cockroach's mouthparts also contain PALPI, which are used quite often to outsmart us humans. These are four little feelers that can taste food without eating it. This means that if a cockroach finds poison that has been set out, it can discover the danger without taking a single bite.

Thorax The THORAX is the middle section of a cockroach's body. Here is where

The powerful legs of this Australian cockroach are attached at the insect's thorax.

the legs and wings are located. Most species of cockroaches have wings. But oddly enough, they rarely fly when trying to escape from an enemy. Instead they usually run.

Cockroaches actually have two pairs of wings. The top pair, called TEGMINA, are longer and stronger than the HIND WINGS, or bottom pair.

Cockroach legs are slender and covered with stiff hairs that the insect uses for touching. On the foot of each leg are two claws. Along with the leg hairs, they help the insect to climb up walls and trees. The legs also give the cockroach its speed—up to 3 miles (4.8 km) per hour. That may not sound very fast, but have you ever tried to stomp a cockroach, only to keep missing it? If so, then you know firsthand how fast 3 miles (4.8 km) per hour can be!

Each of the insect's legs has three knees, which are used to sense vibrations made by possible enemies. In fact, the knees are so sensitive they can actually feel the movement of another insect!

Even though cockroach wings are long and strong, the insect rarely flies.

Bringin' up the Rear

On the rear end of a cockroach are two little hairs called CERCI. They act as motion detectors and alert the cockroach to the slightest breeze or movement. The male cockroach has pointed parts on its back end called STYLES. The styles help it to mate with a female. Baby cockroaches also have styles, but they disappear on the females before they become adults.

Abdomen The ABDOMEN is the largest section of a cockroach's body. Just under the wings, on top of the abdomen, are several hard, overlapping sections, or plates. These are called TERGITES. The plates on the underside are called STERNITES. These sections give the roach a body that looks like a suit of armor.

Inside the Cockroach The insides of a cockroach are not all that different from the insides of humans. Like us, they have a brain, a heart, and saliva glands. That's right! Cockroaches can spit!

The brain of a cockroach is unique. It's not just

in the head, but spread throughout the body of the insect. So, if a cockroach gets its head chopped off, it can still function for up to a week! It finally dies because it cannot eat or drink.

Cockroach Guts A cockroach heart is not shaped like a human's but looks more like a tube with valves. It pumps blood, which is clear or whitish in color, both forward and backward through the body. What's more, the heart can stop, without harming the insect at all.

Have you ever stomped on a cockroach and seen all the white squishy stuff that squirts out? That's body fat, which it uses

This colorful Cuban cockroach (Panchlora nivea) *proves that not all roaches are black or brown.*

for energy and nutrients. Another bonus of body fat is that it helps to detoxify poisons that are sprayed on the insects. Maybe that's why cockroaches are so hard to kill!

How Cockroaches Breathe A cockroach breathes air through tiny holes called SPIRACLES, which are located on top of the thorax and on the sides of the abdomen. They connect directly to the insect's respiratory system, and each spiracle has a little closing device that prevents water from coming in or going out.

Did you know that cockroaches can hold their breath for more than 15 minutes? That helps them get into our homes—by swimming up sewer and water pipes!

A plant leaf is the perfect resting spot in a hot jungle.

This Tropical cockroach is just as pretty as the flowers it sits on.

It's a Roach's Life

As you can imagine, cockroaches reproduce quickly. The female cockroach can produce anywhere from 15 to 50 babies in one hatching, depending on the species. She will mate only once in her lifetime, but will stay fertilized until she dies. This allows her to keep reproducing without the help of a male. It's no wonder there are so many cockroaches in the world!

A Tropical cockroach inspects a leaf while on its search for a mate.

A cockroach goes through three development stages in its lifetime. The stages are egg, nymph, and adult. Of course, the cycle all begins with mating.

Mating The female attracts a mate by releasing pheromones into the air. When the male catches the scent, he answers by raising and fluttering his wings.

He will search for the female until he finally finds her. Once they are together, the male holds on to the female and positions himself underneath her. If the mating is successful, they end up changing positions until they are end to end, facing away from each other. They need to stay like this for about an hour, or until fertilization is complete.

Egg Stage It takes only a few days after mating for the female to develop an egg sac, called an OOTHECA. Depending on the species, the female will either carry the ootheca around on her back or drop it in a hiding place. The eggs need to develop for 3 to 12 weeks. When the babies are ready to hatch, the ootheca simply splits open and out they come.

Many female cockroaches like to drop their egg sacs in hiding places, such as dresser drawers or behind kitchen cabinets.

Nymph Stage Newly hatched cockroaches are called NYMPHS. They look exactly like small versions of the adults, except they don't have wings and are white in color with black eyes. After they are exposed to the air, the bodies of nymphs harden and turn brown.

As the nymph grows, its outer covering, called a CUTICLE, does not grow with it. Instead the nymph MOLTS, or sheds its skin. When the time is right, the cuticle splits open and a slightly larger insect crawls out wearing a new skin. If a nymph loses a body part like a leg or antenna, it can grow a new one between molts. When it crawls out of its

old skin, it will be a new, whole insect!

This nymph stage can last from 5 to 15 months, depending on the type of cockroach. The nymph molts 6 to 13 times before it reaches adulthood.

Adult Stage A roach's life expectancy is anywhere from two to four years—providing it doesn't get squashed by a shoe or eaten by an animal. Most adult cockroaches are NOCTURNAL, and come out only at night to feed. During the day, they rest. In fact, a cockroach spends 75 percent of its time (or 18 hours a day) resting. Now that's lazy!

Hiding Places Cockroaches like dark places, such as cracks and crevices. Some live in jungles under piles of leaves and rotting logs to help protect them from

Cracks like these are a good hiding place for this Cuban cockroach.

These Australian cockroaches range in size from tiny nymphs to young adults.

Nasty Eating Habits

Cockroaches are disgusting eaters. They eat just about anything: food, glue, books, other cockroaches, and even animal feces (waste)! Some of their favorite foods are white bread and boiled potatoes.

If a roach doesn't find food during the night, it will not keep looking after the sun comes up. Instead, it waits until the next night to search again—even if it's still hungry! It can actually go two to three weeks without food, but not more than a day or two without water. These insects need liquid just as much as humans do.

predators. They like warmer temperatures, usually between 52 and 95 degrees Fahrenheit (11 and 35 degrees Celsius), and places where water is easily available. A cockroach can withstand temperatures as low as 32 degrees Fahrenheit (0 degrees Celsius); below that it will freeze to death.

Unfortunately, the places where we eat, like kitchens and restaurants, also make ideal homes for cockroaches because they offer warmth, food, and places to hide.

Cockroaches Can Spread Disease Because of their bad eating habits, cockroaches leave a bad smell and spread germs wherever they go. Imagine a cockroach eating animal droppings, then helping itself to crumbs left on the kitchen counter. That's one nasty way it spreads germs!

Some people even develop cockroach allergies. The tiny feces of German cockroaches, for example, can become airborne with normal household dust. Guess what you're breathing in if this happens!

Cockroaches eat just about anything, including plant leaves.

Battling Roaches

Let's face it, most people do not like living with cockroaches. If you think you may have roaches, or just want to prevent them from moving in, there are some steps you can take to make life hard for them.

Patch and Plug Cockroaches like to crawl up drains and into cracks. An adult male needs a space only as thick as a quarter to squeeze through. To stop this from happening, you can fill cracks with caulking

An open drain can be an open invitation for a cockroach visitor.

This American cockroach would like nothing more than to move in with humans.

and keep the sinks plugged. You can also pour a disinfectant cleaner down the drain every night. This will keep them from coming up the drain pipes.

Clean and Dry A good way to keep the bugs out is to take away their food supply. Make sure you vacuum and sweep up after meals to remove any crumbs that may have fallen on the floor, especially under the stove and refrigerator. Always keep opened food stored in airtight plastic containers. If you have a pet, clean up after it eats to remove all traces of food. That goes for you, too! Always make sure your dishes have been cleaned with hot, soapy water.

Water is more important to cockroaches than food. Leaky faucets, half-full glasses,

The Tropical cockroach enjoys the constant moisture in its rainforest habitat.

and even water in the saucers of potted plants are invitations for a cockroach invasion. If you take away the water, you may not get as many six-legged visitors.

Insect poisons and sprays are also a good way to fight these pesky insects. However, these should be used only by adults. Many can cause harm to humans and pets, if not used correctly.

The different species of cockroaches are scattered all over the world, thanks to humans. Many times, travelers accidentally bring the insects along with them in their luggage. Others may stow away aboard a ship full of food or other goods. But not all roaches are world travelers. Here are just a few of the many species of cockroaches you might see in your own neighborhood.

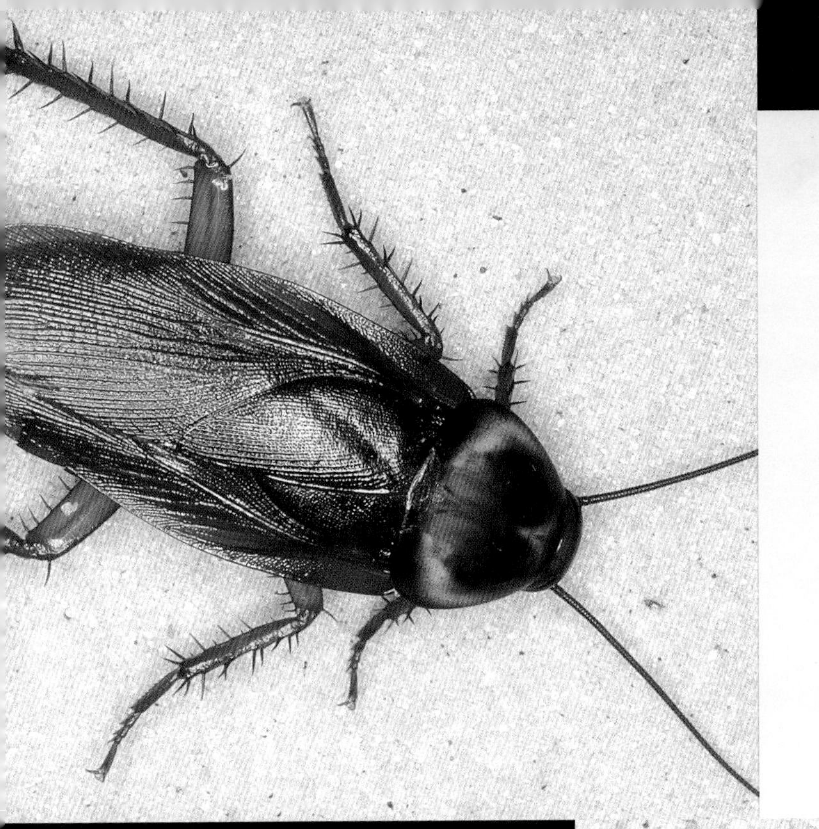

American Cockroach

The American cockroach *(Periplaneta americana)* is very large. It measures from 1 to 2 inches (2.5 to 5 cm) long and has a reddish-brown body with a little yellow on its thorax. Two of the other names for this cockroach are "water bug" and "palmetto bug." It prefers a warm, moist environment. Basements, restaurants, and grocery stores are some places that are commonly invaded by this insect.

German Cockroach

The German cockroach *(Blattella germanica)* is the smallest on the "pest" list. It ranges in length from 1/2 inch to 5/8 inch (1.3 to 1.5 cm) and can be recognized by its tan color and the two black streaks that run down its back. Favorite hiding places of this little pest are under kitchen appliances, around water pipes, and in drawers and cabinets.

Australian Cockroach

The Australian cockroach *(Periplaneta australasiae)* is similar to the American cockroach, only it's a little smaller. Florida is a very popular state for this insect to live in because it is very warm. Greenhouses are another favorite hiding place for cockroaches.

Oriental Cockroach

The Oriental cockroach *(Blatta orientalis)* is reddish-brown or black in color, which is why it's sometimes called the "black beetle." It can range in size from 5/8 inch to 1-1/4 inches (1.3 to 3.1 cm). This species prefers temperatures a little cooler than its other relatives do, and unlike the others, it cannot fly. Most often it can be found in basements, bathrooms, and sewers.

Cockroaches as Pets?

That's right! Some people keep cockroaches as pets! A favorite of roach-lovers is the Madagascar Hissing Cockroach (*Gromphadorhina portentosa*). This insect is a whopping 2 to 3 inches (5 to 7.5 cm) long, and is native to Madagascar, an island off the east coast of Africa. It gets its name from the hissing sound it makes when threatened. The sound, which comes from air being forced through its spiracles, resembles the hiss of a snake. Wouldn't you run if a cockroach hissed at you?

The Madagascar Hissing Cockroach is big and noisy, for a bug, but they are perfectly harmless.

While you might think a cockroach would make a neat pet, Mom or Dad may have a different idea. It would probably be best to ask first. If everyone is agreed that a cockroach is okay, then you will need a few things to raise it properly.

Cockroach Home First on the list is an escape-proof cage, such as a TERRARIUM, or dry aquarium. It needs a tight-fitting mesh lid so that air can get in; place it in a warm room. Make sure you provide bedding, like wood shavings, and hiding places for the insect. Empty paper towel rolls and small boxes work well.

Fresh food and water should always be kept available. Remember, cockroaches especially like white bread and boiled potatoes. For a treat you can give them pieces of cinnamon roll! If you get a male and a female, you might even be able to watch the roach life cycle.

One thing is for sure: You will be the talk of the neighborhood!

This Australian cockroach would make a great pet—and you may be able to catch one in your own backyard!

LEARNING RESOURCES

BOOKS

Cockroaches, Joanna Cole, Morrow Publishing, 1971

Cockroaches, Tamara Green, Gareth Stevens Publishers, 1997

Cockroaches, Mona Kerby, Franklin Watts Publishers, 1989

Cockroaches, Lynn M. Stone, Rourke Book Company, 1995

Cockroaches: Here, There, and Everywhere, Laurence P. Pringle, Crowell Publishing, 1971

The Compleat Cockroach, David G. Gordon, Ten Speed Press, 1996

CHAPTERS IN BOOKS

The Big Bug Book, "Madagascar Hissing Cockroach," Margery Facklam, Little, Brown and Company, 1994, p. 20

Insect Pests, Fitcher and Zim, Golden Press, 1966, pp. 28–29

Insects, Elizabeth Cooper, Steck-Vaughn Publishers, 1990, p. 41

Insects, Alice Fields, Franklin Watts Publishers, 1980, pp. 6, 18

Insects, Herbert Zim and Clarence Cottam, Golden Press, 1991, pp. 23, 152–153

FIELD GUIDES

The Amateur Naturalist, Gerald Durrell (naturalist), 1986

Discovering the Outdoors, a Nature and Science Guide, American Museum of Natural History, 1969

Eyewitness–Living Earth, Miranda Smith, DK Publishers, 1996

Insects, Steve Parker, Dorling Kindersley, 1994, pp 18, 29, 34

WEB

"Blattodea," Insect Compendium Index, 1997

"Bug Club Home Page," Amateur Entomologists Society

"The Cockroach Home Page," Biology Department, University of Massachusetts, 1997

"Cockroach Picture Gallery," Pesticide Education Resources, University of Nebraska-Lincoln

"Cockroaches Home Page," Blattodea Culture Group, 1997

"Insects Home Page," Gordon Ramel (entomologist), 1997

ENCYCLOPEDIAS

American Academic Encyclopedia, Vol. 5, Grolier Inc., 1997

Compton's Encyclopedia online

The Encyclopedia of Wildlife, "Cockroaches," Castle Books, 1974, p. 43

Grzimek's Animal Life Encyclopedia, Vol. 2, *Insects*, Van Nostrand Reinhold, 1975

Illustrated Wildlife Encyclopedia, Vol. 6, Funk and Wagnalls, 1980

NSA Family Encyclopedia, Vol. 4, "Cockroaches," Standard Education Company, 1992

MAGAZINE ARTICLES

"Cockroach Scent as Status Symbol," *Science News*, September 13, 1997, p. 170

"German Cockroach," *Pest Control*, September 1997, p. 55

"Invasion of the Cockroaches," *National Geographic World*, Issue 263, p. 28

"Pest Control Technicians Do Detective Work," *Pest Control*, December 1997, p. 14

"A Powerful Odor," *New York Times*, September 9, 1997, p. C4

"The Seat of Insect Learning?" *Natural History*, September 1997, p. 58

MUSEUMS

Encounter Center
University State Museum
Lincoln, NE

Smithsonian Institution
Washington, DC

INDEX